D1560318

THE LIBRARY OF THE PLANETS™

NEPTUNE

Luke Thompson

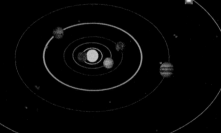

The Rosen Publishing Group's
PowerKids Press™
New York

For Lucy McCann

Published in 2001 by The Rosen Publishing Group, Inc.
29 East 21st Street, New York, NY 10010

Copyright © 2001 by the Rosen Publishing Group, Inc.

All rights reserved. No part of this book may be reproduced in any form without permission in writing from the publisher, except by a reviewer.

First Edition

Book Design: Michael Caroleo and Michael de Guzman

Photo Credits: pp. 1, 4, 11, 22 PhotoDisc; p. 7 (Roman god Neptune) Michael R. Whalen/NGS Image Collection, p. 7 (Neptune) PhotoDisc (digital illustration by Michael de Guzman); p. 8 (Earth) Courtesy of NASA/JPL/California Institute of Technology, p. 8 (Neptune) PhotoDisc (digital illustration by Michael de Guzman); p. 12 (*Voyager*) Courtesy of NASA/JPL/California Institute of Technology, p. 12 (Neptune) Courtesy of NASA/JPL/California Institute of Technology (digital illustration by Michael de Guzman); p. 15 Courtesy of NASA/JPL/California Institute of Technology; p. 16 (Neptune) Courtesy of NASA/JPL/California Institute of Technology, p. 16 (Hubble Space Telescope) © NASA/Roger Ressmeyer/CORBIS (digital illustration by Michael de Guzman); p. 18 Courtesy of NASA/JPL/California Institute of Technology; p. 19 Courtesy of NASA/JPL/California Institute of Technology; p. 20 Courtesy of NASA/JPL/California Institute of Technology.

Thompson, Luke.
  Neptune / Luke Thompson.—1st ed.
    p. cm. — (Library of the planets)
  Summary: Examines the history, unique features, and exploration of Neptune, the eighth planet from the sun.
  ISBN 0-8239-5648-2 (alk. paper)
    1. Neptune (Planet)—Juvenile literature. [1. Neptune (Planet)] I. Title. II. Series.

QB691 .T48 2000
523.48'1—dc21
                                                          00-025039

Manufactured in the United States of America

Contents

1	A Gas Giant	5
2	Discovering Neptune	6
3	Neptune and Earth	9
4	Sister Planets	10
5	*Voyager 2*	13
6	The Storms and Winds of Neptune	14
7	The Hubble Telescope	17
8	Neptune's Moons	18
9	The Rings of Neptune	21
10	Future Visits to Neptune	22
	Glossary	23
	Index	24
	Web Sites	24

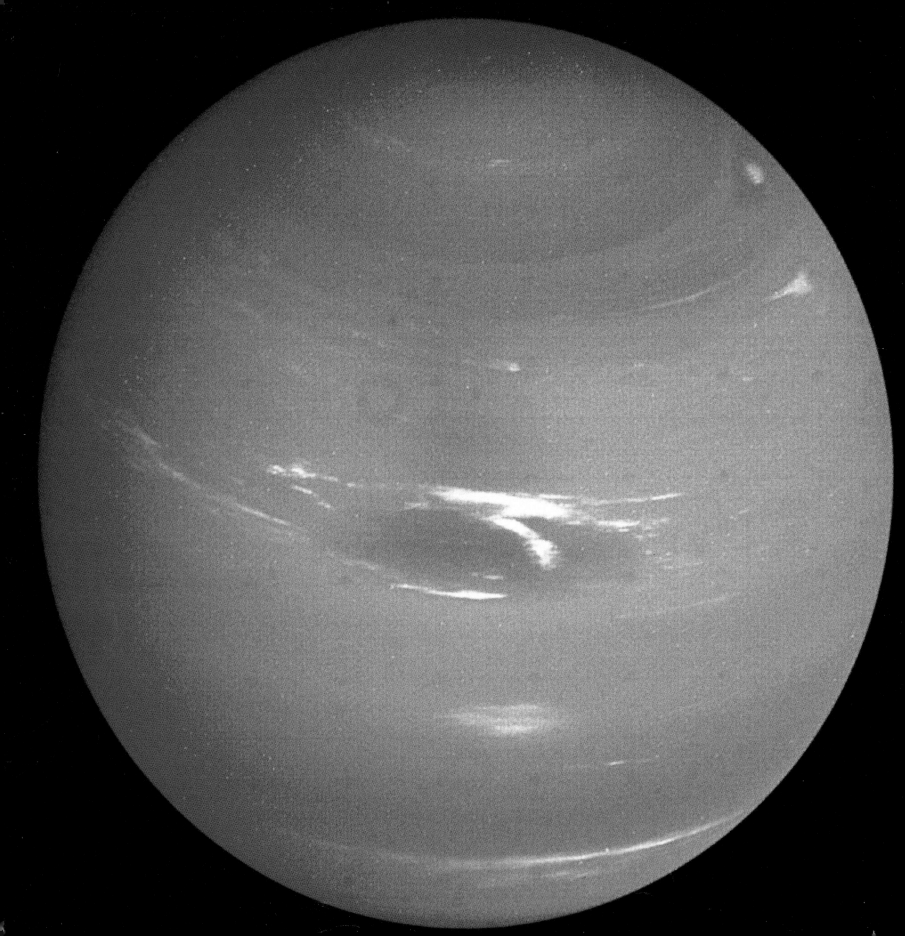

A Gas Giant

The planet Neptune is 31,000 miles (49,890 km) across. This makes it the fourth largest of the nine planets in our **solar system**. A solar system is made up of a group of planets that **orbit** a star. In our solar system that star is the Sun. Neptune is also one of the **gas giants,** which are huge planets made up mostly of gas. The other gas giants are Jupiter, Saturn, and Uranus. These three planets are the first, second, and third largest planets in the solar system.

Neptune orbits the Sun slowly because it is far away from it. Neptune is about 2.8 million miles (4.5 million km) from the Sun. It orbits the Sun once every 165 Earth years. It will not finish its current orbit until 2014.

Neptune is one of the outer planets, which are the planets farthest from the Sun. The other outer planets are Saturn, Uranus, and Pluto.

Discovering Neptune

Like all the planets, Neptune **rotates**, or spins, on its **axis**. An axis is an imaginary pole that runs through the center of a planet. Neptune completes one turn on its axis every 16 hours and 7 minutes. The **rotation** of another planet, Uranus, led to the discovery of Neptune. Uranus was discovered in 1781. A few years later, **astronomers** noticed that Uranus didn't always follow the path that they expected. They thought that another planet must have been pulling on Uranus, changing its orbit. The astronomers used math to figure out where the other planet was. In 1846, Neptune was discovered. It was named after the Roman god of the sea because of its blue color.

Neptune is usually the eighth planet from the Sun. Once every 200 years or so, the planet Pluto travels inside of Neptune's orbit. When this happens, Neptune becomes the planet farthest from the Sun.

Every planet except Earth was named after a god from Greek or Roman mythology. Neptune was named after the Roman god of the sea. He is often shown carrying a trident, or three-pronged fork.

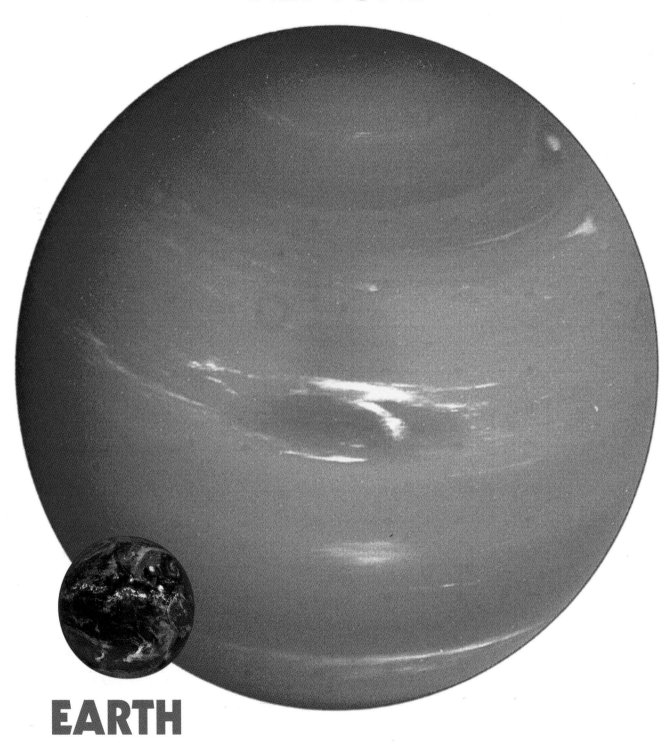

Neptune and Earth

Neptune is much bigger than Earth. In fact if Neptune were hollow, you could fit about 42 Earths inside! Neptune is also much farther from the Sun than Earth. This means that Neptune gets a lot less heat and light from the Sun than Earth does. It gets less than one percent of the sunlight that reaches Earth. The light given off by the Sun reaches Earth in just a few minutes. The same light takes nearly four hours to reach Neptune. The temperature on Neptune can reach as low as -373 degrees Fahrenheit (-225 degrees C). The temperature on Earth rarely goes below -100 degrees Fahrenheit (-73.3 degrees C) even in the coldest parts of the planet.

FUN FACTS

Planet	Orbit Time Around the Sun
Mercury	88 Earth days
Venus	225 Earth days
Earth	1 year (365 days)
Mars	1 year and 322 Earth days
Jupiter	12 Earth years
Saturn	29 Earth years
Uranus	84 Earth years
Neptune	165 Earth years
Pluto	248 Earth years

Neptune's diameter, or the width or thickness across its middle, is almost four times greater than Earth's.

Sister Planets

Neptune is most often compared to the planet Uranus. They are sometimes called sister planets because they have so many things in common. Neptune and Uranus are both pale blue planets. They are close in size, although Neptune is a little smaller. Like Uranus, Neptune does not have a solid surface. The surface of Neptune is made up mostly of two gases, hydrogen and helium, along with water. Scientists think that Neptune has an outer layer of methane, ammonia, and water. They also think it has an inner **core** of melted rocky material. At its center, Neptune's temperature is 800 degrees Fahrenheit (427 degrees C).

A planet's **atmosphere** is the layer of gases that surrounds it. Neptune's atmosphere is a mix of hydrogen, helium, and methane. It also has clouds made of ammonia and other gases, along with ice. Above these clouds are layers of frozen methane. It is the clouds of frozen methane gas that give Neptune its blue color.

Of all the planets in our solar system, Neptune is most like Uranus. They are close in size and the same pale blue color.

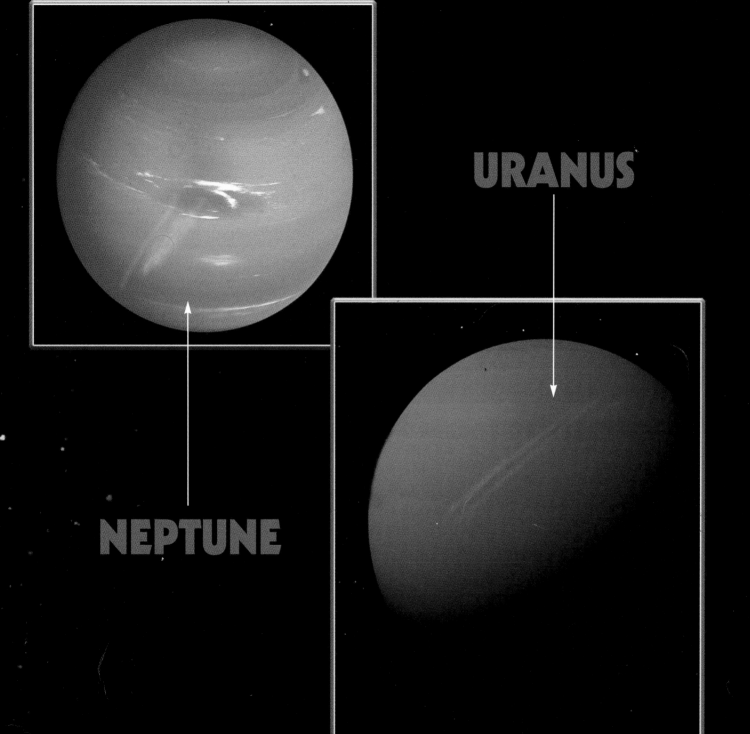

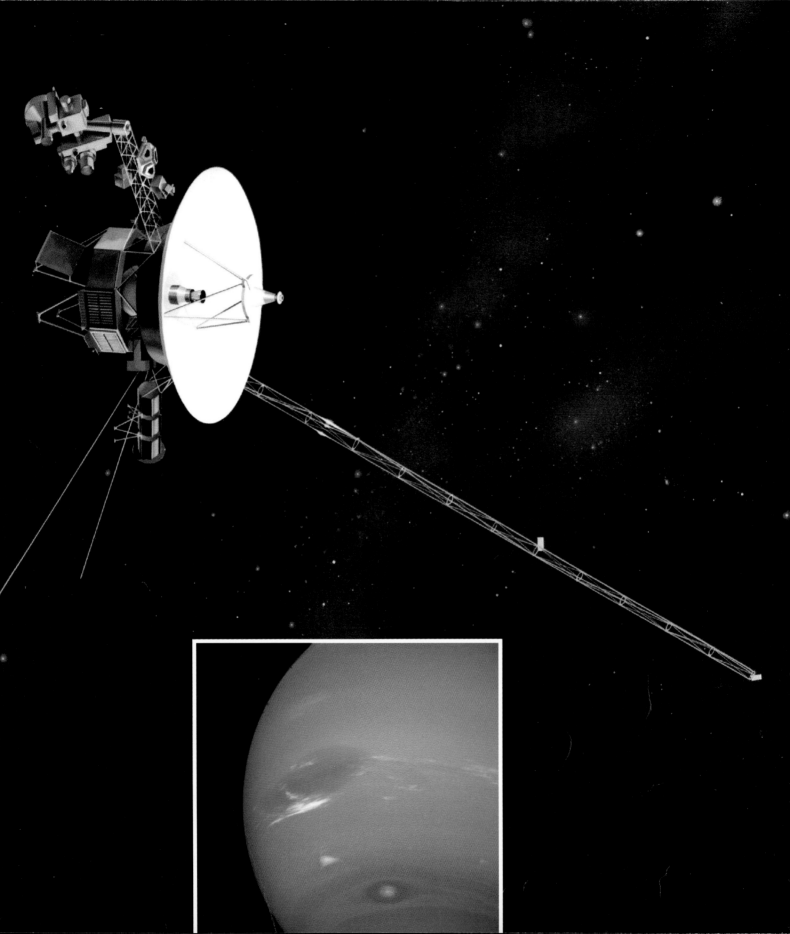

Voyager 2

Two **space probes** were sent out from the United States in 1977. A space probe is a spacecraft that is **launched** without any **astronauts** inside. Space probes are controlled by scientists on Earth. The *Voyager* probes were sent into space to take pictures of the outer planets of our solar system. *Voyager 1* got as far as Saturn before drifting off into space. *Voyager 1* never reached Uranus or Neptune. It took *Voyager 2* more than eight years to get to Uranus. It took another three and a half years to get from Uranus to within 3,000 miles (4,828 km) of Neptune. When *Voyager 2* passed Neptune in 1989, it had been traveling for about 12 years. It had already gone nearly 3 billion miles (4.8 billion km). We learned most of what we now know about Neptune and the other outer planets from the pictures sent back by *Voyager 2*.

The space probe Voyager 2 *(top) took this picture, which shows the Great Dark Spot of Neptune.* Voyager 2 *came within 3,000 miles (4,828 km) of Neptune in August 1989.*

The Storms and Winds of Neptune

Neptune is a planet with violent storms and winds. One storm was so big that it could be seen on the pictures that *Voyager 2* sent back from space. Scientists named this storm the Great Dark Spot. Some scientists believed that the Great Dark Spot was a hole in Neptune's clouds. The Great Dark Spot disappeared in 1994. Soon after, a spot much like it appeared on Neptune's other **hemisphere**.

Neptune's winds are the strongest of any planet in the solar system. *Voyager 2* recorded winds in Neptune's atmosphere moving from 400 miles (644 km) to 1,500 miles (2,414 km) per hour.

Scientists believe that the Great Dark Spot (in box) was as large as Earth's Pacific Ocean. Over the years, it changed its shape. In 1994, it completely disappeared.

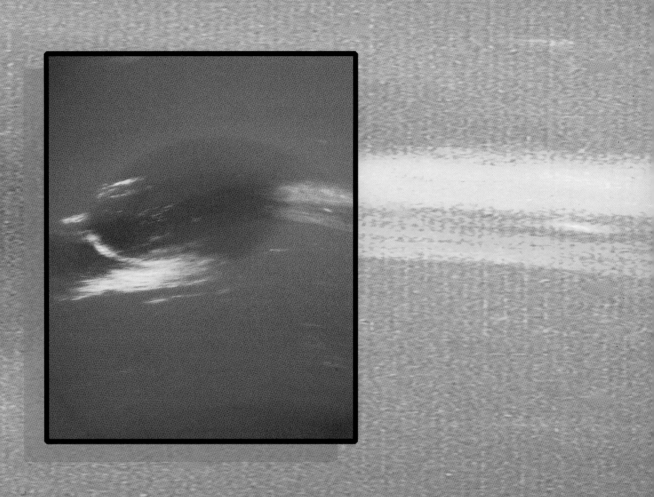

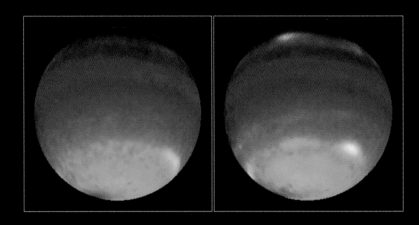

The Hubble Telescope

On April 25, 1990, a very powerful **telescope** was sent into orbit around Earth. This telescope was called the Hubble Space Telescope. The Hubble Telescope sent back its first pictures on May 20, but the pictures were blurry! When scientists fixed the telescope, it started sending back pictures that were clearer and more detailed than any that had been taken in space before. The Hubble Space Telescope showed images of the weather on Neptune's Northern and Southern Hemispheres. Neptune's strong winds were seen as a dark blue belt south of Neptune's **equator**. The Hubble Telescope was also able to track the movement of clouds on Neptune. In 1995, it sent back pictures of a dark spot in the southern part of Neptune. This dark spot showed up after the Great Dark Spot disappeared.

In 1994, the Hubble Space Telescope sent the first close-up views of Neptune since Voyager 2 passed by the planet in 1989. Hubble's images helped track storm changes, like those in the Great Dark Spot.

Neptune's Moons

Before *Voyager 2*, scientists only knew about two of Neptune's moons. Pictures from the space probe showed that the planet had six more. Triton was the first moon to be discovered circling Neptune. Triton is 1,680 miles (2,703 km) wide, about three-quarters as wide as Earth's moon. It is about the same distance from Neptune as our moon is from Earth. Triton is Neptune's biggest moon. Triton's surface temperature is -391 degrees Fahrenheit (-235 degrees C). That is colder than any other part of the solar system that we know about. Pictures from *Voyager 2* showed what

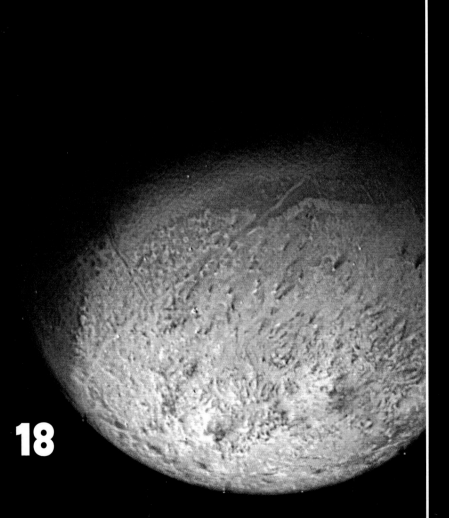

seemed to be **geysers** on Triton. They were actually **explosions** of hydrogen gas that carried tiny bits of dust. The second moon to be discovered orbiting Neptune is called Nereid. Like the six other moons, it is much smaller than Triton. Nereid is the moon that is the farthest away from Neptune.

Neptune's largest moon is Triton. Many astronomers think Triton's surface looks like the skin of a cantaloupe.

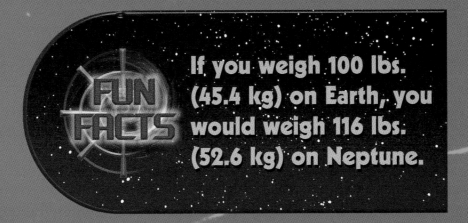

FUN FACTS: If you weigh 100 lbs. (45.4 kg) on Earth, you would weigh 116 lbs. (52.6 kg) on Neptune.

The Rings of Neptune

Voyager 2 also found **rings** around Neptune. These rings are made from **fragments** of rock and ice that form a circle around the planet. Four of the planets in the solar system have rings around them. These planets are Jupiter, Saturn, Neptune, and Uranus.

Neptune's rings are very thin compared to the rings around Jupiter and Saturn. Neptune's rings look lumpy and uneven. Some of the rings that circle Neptune do not go all the way around the planet. Neptune's rings are very dark. They are hard to see from far away. This is because the rings are made out of carbon, which is a very dark material.

Voyager 2 took this picture of Neptune's rings. The outermost ring is about 39,000 miles (63,000 km) away from the planet. The closest ring is about 26,000 miles (42,000 km) above Neptune.

Future Visits to Neptune

Voyager 1 and Voyager 2 are now heading out of the solar system. They should still be sending messages back to Earth for another 20 to 30 years. There is no plan to visit Neptune again. The next space mission planned by the United States is going to be a mission to Pluto, the ninth planet. This does not mean that scientists will stop studying Neptune. The Hubble Space Telescope continues to send back images from outer space. There is always more information to be discovered about all the planets in our huge solar system.

Glossary

astronauts (AS-troh-nots) Members of a crew on a spacecraft.
astronomers (uh-STRAH-nuh-merz) People who study the sun, moon, planets, and stars.
atmosphere (AT-muh-sfeer) The layer of gases that surrounds an object in space. On Earth, this layer is air.
axis (AK-sis) A straight line on which an object turns or seems to turn.
core (KOR) The center layer of a planet.
equator (ih-KWAY-tur) An imaginary line that separates a planet into two parts, North and South.
explosions (ehk-SPLOH-junz) When things blow up.
fragments (FRAG-mints) Small pieces broken off from something else.
gas giants (GAS JY-antz) Planets made up mostly of gas.
geysers (GUY-serz) Springs that send up jets of hot water or steam.
hemisphere (HEM-uh-sfeer) Half of a sphere or globe.
launched (LAWNCHD) Pushed out or put into the air.
orbit (OR-bit) When a planet circles around another object.
rings (RINGS) Thin bands of rock, ice, and other material that stretch around the four giant planets.
rotates (ROW-tayts) When something moves in a circle.
rotation (row-TAY-shun) The spinning motion of a planet around its axis.
solar system (SOH-ler SIS-tem) A group of planets that circles a star. Our solar system has nine planets, which circle the Sun.
space probes (SPAYS PROHBZ) Spacecraft that travel in space and are steered by scientists on the ground.
telescope (TEL-uh-skohp) A device used to make distant objects appear closer and larger.

Index

A
astronomers, 6
axis, 6

E
equator, 17
explosions, 19

G
gas, 5, 10
geysers, 19
Great Dark Spot, the, 14, 17

L
light, 9

M
moons, 18, 19

R
rings, 21

S
solar system, 5, 13, 14, 18, 21, 22
space probes, 13, 18

T
telescope, 17, 22

V
Voyager 1, 13, 22
Voyager 2, 13, 14, 18, 21, 22

If you want to learn more about Neptune, check out these Web sites:
http://pds.jpl.nasa.gov

OCT 1 8 2000
Meherrin Regional Library
133 West Hicks Street
Lawrenceville, VA 23868